This booklet by the Astronomy Correspondent of *The Times*, Michael J. Hendrie, provides a convenient guide to seeing the stars, Moon and planets with the naked eye. The twelve monthly charts on pp. 4–26 show those objects above the horizon late in the evening. The charts have been drawn for the latitude of London (51° 30' north) but may be used for any part of the British Isles. Opposite each chart are notes on the visibility of the planets and phases of the Moon. Notes on the principal meteor showers are on p. 28 with a table of the times of sunset, twilight and sunrise on p. 29. Notes on the stars and the measurement of their brightness are on p. 30. A brief description of the solar system and notes on further events in 2006 are on pp. 31–32. A detailed explanation of the terms used and how the various phenomena arise can be found in the fully illustrated *The Times Night Sky Companion*.

The Changing Aspect of the Night Sky

From our position on the surface of the Earth, the stars appear to lie on the inside of a spherical surface, called the celestial sphere. The stars are so far away their directions appear unchanged when seen from different parts of the Earth's orbit. On the opposite page are diagrams showing the orbits of the planets, as seen from the north pole (above) of the Earth's orbit. The whole of the upper diagram lies within the orbit of Jupiter, shown in the lower diagram. It is not possible to show all the orbits clearly in a single diagram on a page of this size. Round the circumference are the zodiacal constellations through which the Sun and planets appear to move. To save space the diagrams have been overlapped but the same constellations continue round both diagrams.

The large arrows point to the Vernal or Spring Equinox, also known as the First Point of Aries (the point on the celestial sphere where the Sun crosses the equator from S to N). The stars that lie behind the Sun as seen from the Earth in April each year will, by October, be in the opposite part of the sky to the Sun and be due south at midnight. The arrows point to Pisces (where the First Point of Aries now lies) and looking at the October chart Pisces is indeed in the southern sky but does not appear at all on the April chart, being behind the Sun and in the daytime sky. A line from the Sun through the position of the Earth points towards those stars and planets seen near the lower centre of the chart for that month. Remembering this, one can relate the positions of the planets in their orbits to where they will be in the sky, but they will not always appear on the monthly chart if they are too near the Sun in direction or below the horizon at that time.

Time of Observation and Location

Universal Time (UT), which is equivalent used throughout this booklet. British Summ late March to late October and is 1h ahead charts are drawn for an observer near Lon meridian). As one moves north Polaris becon appear above the southern horizon. Movem alter which stars can be seen, only when th

GW00696162

Using the Charts

The charts show the brighter stars above the horizon for London at 23h (11pm) at the beginning, 22h (10pm) in the middle and 21h (9pm) at the end of each month. The stars rise approximately 4 minutes earlier each night or 2 hours earlier each month, being back in the same positions at the same time after a year. Thus for example, the aspect of the heavens at 23h on 1 April is the same as on 1 May at 21h or 1 March at 01h. By remembering this rule, the chart for any hour throughout the year can be found. This rule does not apply to the Moon and planets. The charts show the whole sky visible at one time with the zenith, the point directly overhead, in the centre. Note that the Pole Star (Polaris) occupies the same position on every chart, being close to one of two points around which the whole star sphere appears to revolve. It is easily found in relation to Ursa Major at all times of the year and is useful for finding due north. Ursa Major's seven brightest stars form the Plough, Dipper, Wagon or Wain. The two leading stars (The Pointers) are always in line with Polaris. If the observer faces south with the Pole Star to his back and the appropriate chart held up to the front, the constellations depicted above the southern horizon will be to the front, those to the east on the left and those to the west on the right.

The Times Night Sky Starfinder can be set to show the stars above the local horizon for any time of the night for any year. It does not show the positions of the Moon and planets as these change from year to year, but these are given on the charts in this booklet.

Explanatory Notes on the Terms Used

The Moon phase and positions are given for about 22h on every other day when it is above the horizon at that time. The average interval between like phases (e.g. full to full) is 29.5 days, 2 days longer than it takes to return amongst the same stars. The Moon moves eastwards by its own diameter every hour.

The planets are shown in the position they occupy about the middle of the month unless otherwise indicated, and for Mars an arrow shows by its length the movement during the month. Planets crossing the meridian (i.e. due south) before midnight are called *evening stars* while those crossing after midnight are *morning stars*. A planet is at opposition when it is in the opposite part of the sky to the Sun and therefore due south at midnight. (Mercury and Venus can never be at opposition.) The planet is then at its closest to the Earth and at its brightest for that year. For a few weeks on either side of opposition, the planet's motion against the stars, instead of being direct or from west to east as usual, is from east to west and is called retrograde. At the turning points, where the motion is reversed, the planet is said to be stationary. A planet coming into line with the Earth and Sun is at superior conjunction if it lies beyond the Sun but at inferior conjunction if it lies between the Earth and Sun. Only Mercury and Venus can be at inferior conjunction. The term 'conjunction' is also used to describe planets that appear close to other planets. Mercury and Venus are at greatest elongation when at their greatest apparent distance from the Sun, either east (evening) or west (morning). They can never be high in the sky in the middle of the night.

Mercury is usually in twilight as seen from the British Isles and may require binoculars, though it can be an easy naked eye object when near maximum brightness. Neither Mercury nor Venus appears on the charts in 2006. Uranus (5.7 mag) and Neptune (7.8 mag) are not naked eye objects to the average eye, but are shown for those with access to a telescope. Pluto is 14th magnitude and requires a moderate-sized telescope and detailed charts. It is at opposition in Ophiuchus on 16th June 2006.

Eclipses and Transit of Mercury

14-15 March

This penumbral eclipse of the Moon is wholly visible from the British Isles, Europe and Africa but as the Moon passes only through the Earth's penumbral or outer, lighter shadow it will be only slightly darkened and the eclipse may well pass unnoticed. The Moon enters the penumbra at 21h 21m and leaves at 02h 13m.

29 March

This total eclipse of the Sun begins in the extreme E of Brazil, crosses the N Atlantic to Ghana, Nigeria, Libya and Turkey, Russia, Kazakstan and Mongolia. The eclipse begins about 08h in Ghana where totality lasts about 3.5 minutes. A partial eclipse will be seen over a wide area, including the British Isles where between 10h and 11h a maximum of about 17 per cent of the Sun will be covered by the Moon.

7 September

This partial eclipse of the Moon centred on the Indian sub-continent will not be spectacular, as the Moon only just dips into the umbra. Even at mid-eclipse at 18h 51m only a small dark notch in the Moon will be seen. The Moon rises in the British Isles just before mid-eclipse and leaves the umbra at 19h 37m.

22 September

The track of this annular eclipse of the Sun begins in Guyana and runs SE across the S Atlantic to the Southern Ocean between S Africa and Antarctica. A partial eclipse will be seen over S America, S and W Africa and part of Antarctica.

8-9 November

A transit of Mercury over the face of the Sun will be visible over a wide area centred on the Pacific Ocean including E Asia and Australasia, most of the Americas and part of Antarctica. It will not be visible from W Asia, Europe or Africa. Least separation will occur about 21h 41m.

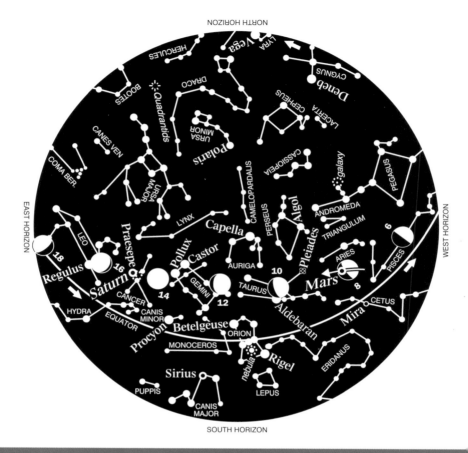

JANUARY 1, 23h (11pm)

The aspect of the sky (apart from the Moon and planets) will be approximately the same in other months at the following times:

October 1, 05h: November 1, 03h: December 1, 01h: February 1, 21h: March 1, 19h.

The time in these notes is that of the Greenwich meridian.

The Planets

MERCURY   is not observable before superior conjunction on 26th, and then moves into the evening sky.

VENUS   visible after sunset in early January, at inferior conjunction on 14th. By 31st rising about 05h 30m in the SE, a brilliant -4.4 magnitude. Moon nearby on 1st.

MARS   in Aries fades from -0.6 to +0.2 mag this month. Setting about 03h 30m on 1st and 02h 30m by 31st. Moon very close above on 8th.

JUPITER   is -1.9 mag, in Libra, rising soon after 03h on the 1st and by 01h 30m on 31st. Moon nearby on 23rd and 24th.

SATURN   is -0.1 mag, in Cancer and at opposition on 27th. Moon nearby on 15th.

URANUS   in Aquarius throughout the year and sets about 19h by 31st.

NEPTUNE   in Capricornus throughout the year and sets about sunset by 31st.

The Moon

First Quarter   6d 19h

Full Moon   14d 10h

Last quarter   22d 15h

New Moon   29d 14h

The Earth

At perihelion, its closest to the Sun, on 4d 15h (147 million km).

Rises below Spica on the 22nd and rises near Antares on the 25th.

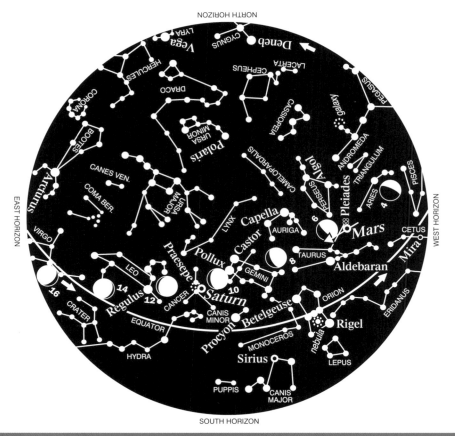

FEBRUARY 1 , 23h (11pm)

The aspect of the sky (apart from the Moon and planets) will be approximately the same in other months at the following times:

November 1, 05h: December 1, 03h: January 1, 01h: March 1, 21h: April 1, 19h.

The time in these notes is that of the Greenwich meridian.

The Planets

MERCURY | in SW twilight from mid-month, setting 2h after the Sun on the 24th, when still -0.3 mag and at greatest E elongation (18°).

VENUS | is a brilliant -4.5 mag morning object, rising by 04h 30m on 28th. Stationary on 3rd.

MARS | fades to 0.7 mag, moving from Aries into Taurus, below Pleiades in mid-month. Moon close above on 5th.

JUPITER | in Libra, brightens to -2.2 mag, rising by 23h 30m by 28th. Moon nearby on 20th.

SATURN | is -0.1 mag, in Cancer, not setting until 06h on 28th. Moon nearby on 11th. Saturn passes through the Praesepe star cluster in early February.

URANUS | is too near the Sun for observation.

NEPTUNE | in conjunction with the Sun on 6th and is not observable.

The Moon

	First quarter	5d 06h
	Full Moon	13d 05h
	Last quarter	21d 07h
	New Moon	28d 01h

Near Spica on the 18th, closest at 04h and near Antares on the 22nd.

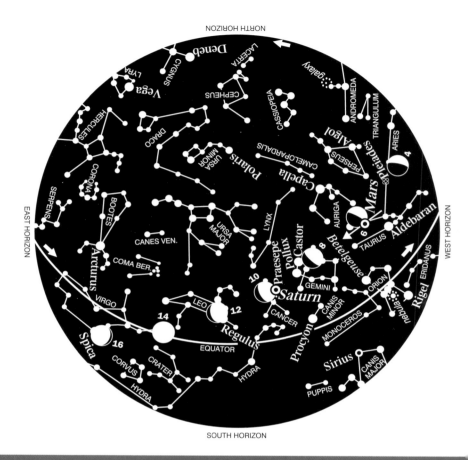

MARCH 1, 23h (11pm)

The aspect of the sky (apart from the Moon and planets) will be approximately the same in other months at the following times:

December 1, 05h: January 1, 03h: February 1, 01h: April 1, 21h.

The time in these notes is that of the Greenwich meridian.

The Planets

MERCURY | at inferior conjunction on 12th, but not visible as a morning object. Next observable on late May evenings.

VENUS | remains in the morning sky until October but mostly in twilight. By 31st rises 1.5 hrs before the Sun, still a brilliant -4.3 mag. Greatest W elongation (47°) on 25th. Moon nearby on 25th.

MARS | is 1.0 mag, in Taurus, setting about 01h by 31st. Moon nearby on 5th.

JUPITER | in Libra, a bright -2.4 mag by 31st when it rises about 21h. Stationary on 5th. Moon nearby on 19th.

SATURN | is 0.0 mag and in Cancer, setting about 04h by 31st. Moon above on 10th.

URANUS | in conjunction with the Sun on 1st, not observable in March.

NEPTUNE | is in morning twilight throughout the month.

The Moon

First quarter 6d 20h

Full Moon 15d 00h

Last quarter 22d 19h

New Moon 29d 10h

Close below Antares about 04h on the 21st.

The Earth

The Spring equinox, when the Sun crosses the equator into the northern hemisphere, on 20d 18h.

Eclipses

of the Moon 14th–15th and the Sun on the 29th (see p. 3)

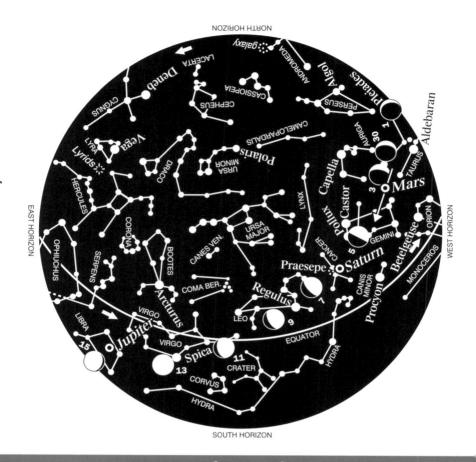

APRIL 1, 23h (11pm)

The aspect of the sky (apart from the Moon and planets) will be approximately the same in other months at the following times:

December 1, 07h: January 1, 05h: February 1, 03h:
March 1, 01h: May 1, 21h.

The time in these notes is that of the Greenwich meridian.

The Planets

MERCURY	rises only 30 mins before the Sun, not visible in April. At greatest W elongation (28°) on 8th.
VENUS	at -4.2 mag, rises only 1 hr before the Sun but should be visible in the E in bright morning twilight. Moon nearby on 24th.
MARS	passes from Taurus into Gemini fading from 1.2 to 1.5 mag during April, setting by 0h 30m by 30th. Moon close above on 3rd.
JUPITER	is -2.5 mag, in Libra, rising about sunset by 30th. Moon nearby on 15th.
SATURN	is 0.2 mag, in Cancer, setting about 02h by 30th. Stationary on 5th. Moon above on 6th–7th.
URANUS	in Aquarius rising about 03h by 30th. Moon nearby on 24th.
NEPTUNE	in Capricornus, rising about 02h by 30th. Moon nearby on 22nd.

The Moon

◐	First quarter	5d 12h
○	Full Moon	13d 17h
◑	Last quarter	21d 03h
●	New Moon	27d 20h

Below Spica on the 13th and near Antares on the 17th.

11

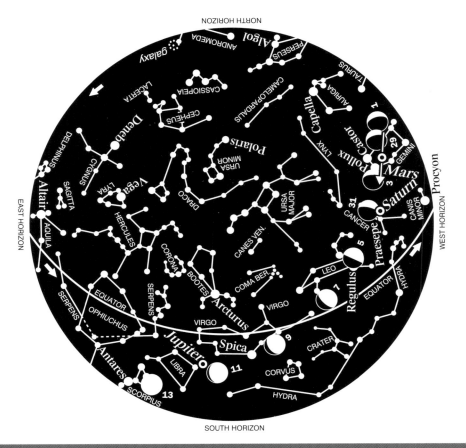

MAY 1, 23h (11pm)

The aspect of the sky (apart from the Moon and planets) will be approximately the same in other months at the following times:

January 1, 07h: February 1, 05h: March 1, 03h:
April 1, 01h: June 1, 21h.

The time in these notes is that of the Greenwich meridian.

The Planets

MERCURY in superior conjunction on 18th, moving into the evening sky. In late May it fades from -1.8 to -1.0 mag, visible in the NW, setting after 21h. Should be a naked eye object in twilight into early June. Moon above on 28th.

VENUS remains in morning twilight, but should be observable in the E before sunrise. Moon above on 23rd.

MARS is 1.6 mag, moving from Gemini into Cancer, setting before midnight by 31st. Moon nearby on 1st–2nd and close above on 30th.

JUPITER is a bright -2.5 mag and in Libra. At opposition on 4th, becoming an evening object. Moon below on 11th–12th.

SATURN is 0.3 mag, in Cancer setting before 0h by the 31st. Moon above on 4th and 31st.

URANUS in Aquarius rises at 01h by 31st. Moon nearby on 21st.

NEPTUNE in Capricornus rises about 0h by 31st, stationary on 22nd. Moon nearby on 19th.

The Moon

◑	**First quarter**	5d 05h
○	**Full Moon**	13d 07h
◐	**Last quarter**	20d 09h
●	**New Moon**	27d 05h

Close below Spica on the 10th–11th.

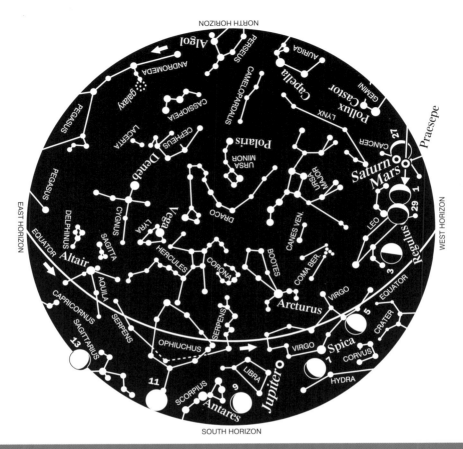

JUNE 1, 23h (11pm)

The aspect of the sky (apart from the Moon and planets) will be approximately the same in other months at the following times:

February 1, 07h: March 1, 05h: April 1, 03h:
May 1, 01h: July 1, 21h.

The time in these notes is that of the Greenwich meridian.

The Planets

MERCURY | fades from -1.0 to + 1.5 mag from 1st to 30th. Setting 2 hrs after the Sun in early June and visible low in NW. Greatest E elongation (25°) on 20th.

VENUS | at -3.9 mag remains in morning twilight but just visible low in ENE Moon above on 23rd.

MARS | is a rather faint 1.7 mag, in Cancer passing through Praesepe star cluster in mid-month, and setting about 22h by 30th. Moon nearby on 28th. Very close below Saturn on 17th.

JUPITER | is -2.4 mag, in Libra, setting about 01h by 30th. Moon below on 8th.

SATURN | is 0.4 mag, passes through Praesepe in Cancer about 4th, setting about 22h by 30th. Moon nearby on 28th.

URANUS | rises about 23h by 30th. Stationary on 19th. Moon nearby on 17–18th.

NEPTUNE | rises about 22h by 30th. Moon nearby on 15–16th.

The Moon

First quarter 3d 23h

Full Moon 11d 18h

Last quarter 18d 14h

New Moon 25d 16h

The Earth

The summer solstice, when the Sun reaches its most northerly point over the Tropic of Cancer, is on 21d 12h.

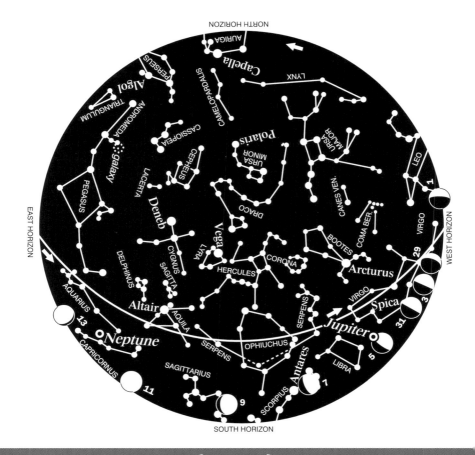

JULY 1, 23h (11pm)

The aspect of the sky (apart from the Moon and planets) will be approximately the same in other months at the following times:

April 1, 05h: May 1, 03h: June 1, 01h:
August 1, 21h: September 1, 19h.

The time in these notes is that of the Greenwich meridian.

The Planets

MERCURY	at inferior conjunction on 18th to become a morning object in August.
VENUS	is a brilliant -3.9 mag rising about 02h through July. Moon nearby on 23rd.
MARS	is 1.8 mag moving from Cancer into Leo, setting 1 hr after the Sun by 31st. Regulus close on 21st–22nd.
JUPITER	is -2.2 mag and in Libra, setting before 23h by the 31st. Stationary and above the Moon on 5th.
SATURN	is 0.4 mag and in Cancer setting about 22h on 1st and soon after sunset by 31st.
URANUS	in Aquarius rising about 21h by 31st. Moon nearby on 14th.
NEPTUNE	in Capricornus rising about sunset by 31st. Moon nearby on 13th.

The Moon

◖	First quarter	3d 17h
○	Full Moon	11d 03h
◑	Last quarter	17d 19h
●	New Moon	25d 05h

Near Spica on 4th and 31st and Antares on 7th–8th.

The Earth

At aphelion, its farthest from the Sun (152 million km), 3d 23h.

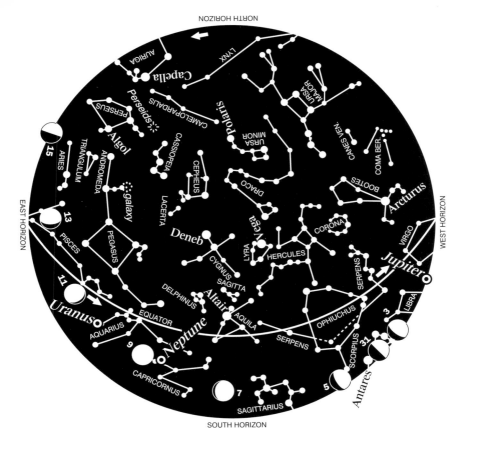

AUGUST 1, 23h (11pm)

The aspect of the sky (apart from the Moon and planets) will be approximately the same in other months at the following times:

June 1, 03h: July 1, 01h: September 1, 21h:
October 1, 19h: November 1, 17h.

The time in these notes is that of the Greenwich meridian.

18

The Planets

MERCURY | at greatest W elongation (19°) on 7th at 0.0 mag, brightening to -1.0 by 15th. It should be visible in NE morning twilight in mid-August. Close below Venus on 9th and near Saturn on 21st.

VENUS | is -3.9 mag, rising 2h before the Sun on 1st. It runs into twilight in September. Moon close by on 22nd.

MARS | sets only minutes after the Sun and is not observable in August.

JUPITER | is -2.0 mag and in Libra, setting about 21h by 31st. Moon nearby on 1st–2nd and 29th.

SATURN | is in conjunction with the Sun on 7th but may be visible close to Venus 26th–28th.

URANUS | rises about sunset by 31st. Moon nearby 10th.

NEPTUNE | is at opposition on 11th and due south at 0h. Moon nearby on 9th.

The Moon

First quarter   2d 09h

Full Moon   9d 11h

Last quarter   16d 02h

New Moon   23d 19h

First quarter   31d 23h

Close below Antares on 4th.

SEPTEMBER 1, 23h (11pm)

The aspect of the sky (apart from the Moon and planets) will be approximately the same in other months at the following times:

**July 1, 03h: August 1, 01h: October 1, 21h:
November 1, 19h: December 1, 17h.**

The time in these notes is that of the Greenwich meridian.

The Planets

MERCURY | in superior conjunction on 1st, then an evening object but too near Sun to be observable.

VENUS | is -3.9 mag in morning twilight, rising only 1 hr before the Sun by 30th. Very near Regulus on 5th–6th.

MARS | sets only minutes after the Sun and is not observable.

JUPITER | is -1.8 mag, in Libra, setting about 19h by 30th. Moon below on 26th.

SATURN | is 0.5 mag, in Leo, rising about 01h 30m by 30th. Moon very close on 19th.

URANUS | at opposition 5th, and due S at 0h. Moon nearby on 7th.

NEPTUNE | sets about 01h by 30th. Moon nearby on 5th.

The Moon

○	Full Moon	7d 19h
◑	Last quarter	14d 11h
●	New Moon	22d 12h
◑	First quarter	30d 11h

Near Antares on 1st and 28th.

The Earth

The autumn equinox, when the Sun crosses the equator into the southern hemisphere, is on 23rd at 04h.

Eclipses

of the Moon on the 7th and the Sun on the 22nd (see p. 3).

OCTOBER 1, 23h (11pm)

The aspect of the sky (apart from the Moon and planets) will be approximately the same in other months at the following times:

August 1, 03h: September 1, 01h: November 1, 21h:
December 1, 19h: January 1, 17h.

The time in these notes is that of the Greenwich meridian.

The Planets

MERCURY | at greatest E elongation (25°) on 17th but too far S of the Sun to be seen.

VENUS | rises only 1hr before the Sun on 1st, in superior conjunction with the Sun on 27th. Just visible before sunrise in early October.

MARS | in conjunction with the Sun 23rd and not visible until December.

JUPITER | is -1.7 mag, in Libra and setting in SW only 30 minutes after the Sun by 31st.

SATURN | is 0.5 mag, in Leo, rising before midnight 31st. Moon above on 16th.

URANUS | in Aquarius setting about 01h 30m by 31st. Moon nearby on 4th.

NEPTUNE | in Capricornus setting 23h by 31st, stationary on 29th. Moon nearby on 2nd.

The Moon

Full Moon   7d 03h

Last quarter   14d 00h

New Moon   22d 05h

First quarter   29d 21h

Below Antares on the 25th.

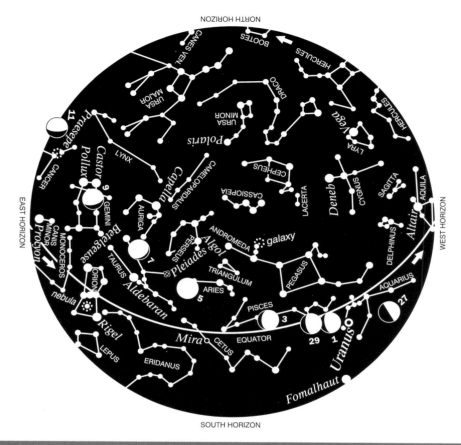

NOVEMBER 1, 23h (11pm)

The aspect of the sky (apart from the Moon and planets) will be approximately the same in other months at the following times:

September 1, 03h: October 1, 01h: December 1, 21h: January 1, 19h: February 1, 17h.

The time in these notes is that of the Greenwich meridian.

The Planets

MERCURY | in inferior conjunction on 8th and moves into the morning sky, brightening to -0.6 mag by 30th, rising 2h before the Sun. At greatest W elongation (20°) on 25th. An easy object in SE twilight from 06h in late November.

VENUS | is too near the Sun for observation.

MARS | in morning twilight and not visible this month.

JUPITER | in conjunction with the Sun on 21st and not observable.

SATURN | is 0.5 mag, in Leo rising by 22h on 30th. Moon nearby on 13th.

URANUS | sets about 23h by 30th. Stationary on 20th. Moon nearby on 1st.

NEPTUNE | sets about 21h by 30th. Moon nearby on 26th.

The Moon

	Full Moon	5d 13h
	Last quarter	12d 18h
	New Moon	20d 22h
	First quarter	28d 06h

Near Spica on the 18th.

Transit of Mercury

The transit of Mercury over the Sun's face on 8d 22h (see p. 3).

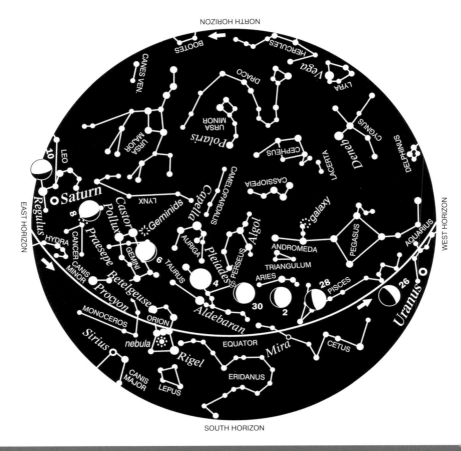

DECEMBER 1, 23h (11pm)

The aspect of the sky (apart from the Moon and planets) will be approximately the same in other months at the following times:

**September 1, 05h: October 1, 03h: November 1, 01h:
January 1, 21h: February 1, 19h.**

The time in these notes is that of the Greenwich meridian.

The Planets

MERCURY | is -0.6 mag during the first 2 weeks of December, visible low in the SE at dawn. Very close to Mars and Jupiter on 8th–12th (see p. 32).

VENUS | visible low in the SW after sunset during late December.

MARS | is 1.5 mag, low in SE morning twilight, moving from Libra into Scorpius. Near Antares on 19th. Moon nearby 19th.

JUPITER | is -1.7 mag moving from Libra, through Scorpius into Ophiuchus, and into a dark sky by 31st. Moon nearby on 19th.

SATURN | is 0.3 mag, in Leo, rising before 20h by 31st. Stationary on 6th. Moon nearby on 9th.

URANUS | sets about 21h 30m by 31st. Moon nearby on 25th.

NEPTUNE | sets about 19h by 31st. Moon nearby on 23rd.

The Moon

◯	**Full Moon**	5d 00h
◑	**Last quarter**	12d 15h
⬤	**New Moon**	20d 14h
◐	**First quarter**	27d 15h

Near Spica on the 15th.

The Earth

The winter solstice, when the Sun reaches its most southerly point over the Tropic of Capricorn is on 22d 00h.

PRINCIPAL METEOR SHOWERS IN 2006

Name	Period of max. activity	Av. hourly rate	Visibility and moonlight FQ/LQ = first/last quarter
Quadrantids	2–4 January	10	Favourable, FQ on 6th. Radiant low N in evening
Lyrids	21–22 April	10	Quite favourable, LQ on 20th
Perseids	11–14 August	60	Unfavourable, full on 9th
Orionids	20–22 October	10–20	Favourable, new on 22nd
Taurids	Late Oct–late Nov	5–10	Slow meteors, from below Pleiades
Leonids	16–18 November	30?	Favourable, new on 20th
Geminids	12–14 December	60	Fairly favourable, LQ on 12th

Notes:
The Leonids gave very strong showers over the past few years but activity has now declined and will remain low for much of this century. Meteor showers are usually named after the constellation from which the meteors appear to radiate. These radiants are shown on the charts except for the Taurids and Leonids. The Taurids cover a wide area and the Leonid radiant (above Regulus) is below the horizon on the November chart. The number of meteors seen depends on when activity peaks, the altitude of the radiant about that time, the darkness and clarity of the sky and the effect of moonlight. A Moon between first and last quarter greatly affects the number of faint meteors seen.

Notes on the Sunset, Sunrise and Nautical Twilight Table:
1 Times are given in Universal Time (=GMT): when British Summer Time (BST) is in force (usually from the late March to late October) add 1 hour.
2 Nautical Twilight ends when the Sun's true centre reaches a depression of 12° below the horizon. Then it is dark enough to see the brighter stars and planets, and in suburban areas it often gets no darker due to artificial lighting. Nautical twilight begins when it is becoming too light to see these stars. When no time is shown nautical twilight lasts all night.
3 The times given are approximate. They depend on the observer's latitude and longitude. Sunset, sunrise and twilight times will be 4 minutes earlier for every degree of longitude east of the Greenwich meridian and 4 minutes later for every degree west. While London is on the Greenwich meridian, Edinburgh is about 3 degrees west or 12 minutes later.
4 The observer's latitude also affects these times: for example sunset occurs earlier in Edinburgh than London in winter but later in summer. It is not possible to cover more than two regions here, but a reasonable estimate can be made for other parts of the British Isles. The times may be used for any year. More explanation on this can be found in *The Times Night Sky Companion*.

SUNSET, SUNRISE AND NAUTICAL TWILIGHT

London area

Date		Sunset	End NT	Begin NT	Sunrise
Jan	1	16 00	17 20	06 45	08 08
	15	16 20	17 40	06 40	08 00
Feb	1	16 45	18 05	06 25	07 40
	15	17 15	18 30	06 00	07 15
Mar	1	17 35	18 50	05 35	06 50
	15	18 05	19 15	05 00	06 15
Apr	1	18 35	19 50	04 20	05 35
	15	19 00	20 10	03 40	05 00
May	1	19 25	20 50	03 05	04 30
	15	19 45	21 35	02 30	04 05
Jun	1	20 10	22 00	01 55	03 50
	15	20 20	22 29	01 33	03 40
Jul	1	20 25	22 25	01 40	03 45
	15	20 20	22 00	02 05	04 00
Aug	1	20 10	21 30	02 40	04 20
	15	19 50	20 50	03 15	04 45
Sep	1	19 25	20 10	03 50	05 10
	15	18 50	19 30	04 20	05 35
Oct	1	18 15	18 50	04 45	06 00
	15	17 40	18 20	05 10	06 25
Nov	1	17 05	17 50	05 40	06 50
	15	16 35	17 25	06 02	07 20
Dec	1	16 10	17 15	06 25	07 45
	15	15 50	17 13	06 37	08 03
	31	16 00	17 20	06 45	08 08

Edinburgh area

Date		Sunset	End NT	Begin NT	Sunrise
Jan	1	15 45	17 22	07 08	08 44
	15	16 10	17 43	07 00	08 45
Feb	1	16 41	18 10	06 40	08 10
	15	17 16	18 38	06 15	07 38
Mar	1	17 47	19 05	05 40	07 03
	15	18 15	19 35	05 05	06 28
Apr	1	18 49	20 12	04 20	05 45
	15	19 20	20 52	03 45	05 05
May	1	19 50	21 35	02 45	04 25
	15	20 15	22 25	01 56	03 47
Jun	1	20 43	23 53	00 35	03 34
	15	21 00	—	—	03 25
Jul	1	21 01	—	—	03 31
	15	20 45	23 40	00 45	03 45
Aug	1	20 19	22 22	02 10	04 15
	15	19 50	21 30	03 00	04 42
Sep	1	19 09	20 42	03 45	05 15
	15	18 30	19 52	04 20	05 43
Oct	1	17 47	19 07	04 55	06 15
	15	17 05	18 32	05 25	06 40
Nov	1	16 31	17 57	05 52	07 19
	15	16 04	17 30	06 22	07 48
Dec	1	15 42	17 16	06 45	08 20
	15	15 36	17 10	07 00	08 38
	31	15 45	17 22	07 08	08 44

(Times in UT)

29

THE DISTRIBUTION OF THE STARS

An examination of the sky on a clear dark night shows that the distribution of stars is far from uniform. While there are distinct clusters of stars like the Pleiades and Praesepe many other groupings consist of stars that just happen to lie in the same direction but at different distances. Binoculars or a small telescope show that the greatest concentration of stars is towards the Milky Way (the faint band of light that passes through Gemini, Auriga, Perseus, Cassiopeia, Cygnus, Aquila and Sagittarius) though not all of these constellations are above our horizon at any one time. The Milky Way extends in a continuous band through the southern constellations that never rise in the British Isles.

Our Sun is situated well away from the centre of a huge, flattened, disc-like system of stars 100,000 light years across called the Galaxy. It contains more than 100,000 million stars. When we look along the plane of the disc we see the star-clouds of the Milky Way, but when we look out above or below this plane we see far fewer stars. From a distance of 2 million light years our Galaxy would look rather like M31 in Andromeda, visible to the naked eye as a hazy patch of light. Other galaxies have been found in their millions, some more than 10,000 million light years away. The central bulge of our Galaxy is towards Sagittarius.

The Brightness or Magnitude of Stars and Planets

The stars (and planets) are subdivided into magnitudes (often called just mag) according to their apparent brightness: the lower the number the brighter the star and the larger the dots on our monthly maps. Any star is about 2.5 times as bright as one of the next magnitude. The faintest visible to the naked eye on a clear, dark night is reckoned to be 6.0 mag, or just one-hundredth the brightness of a 1.0 mag star. The faintest objects now observed are about 30 mag by the Hubble Space Telescope.

Zero magnitude (0.0) represents a brightness 2.5 times that of a 1 mag star, while even brighter objects have a minus sign. Sirius, the brightest star, is -1.47 mag, while the full Moon is -12.5 mag. The magnitudes of selected bright stars are: Polaris (2.0), Aldebaran (0.9), Castor (1.6), Procyon (0.4), Regulus (1.3), Arcturus (0.0), Spica (1.0). Antares 1.0) and Altair (0.8).

In the monthly notes the magnitudes of planets are given as a guide to how easily they may be seen and recognised. Mercury is visible in twilight only, except from equatorial and southern latitudes. It can be as bright as -2 mag near full phase, or as faint as +4 mag when a narrow crescent. Venus varies only between about -3.8 and -4.6 mag and is always very bright. Mars varies from -2.8 to +1.8 mag, while Jupiter varies between -2.8 and -1.7 mag. Saturn varies from -0.5 to +1.0 mag. The distances from the Sun and Earth affect the brightness of all the planets, but the phase has a big effect on Mercury and Venus, while the tilt of the rings is an additional factor for Saturn. Uranus is 5.7 mag, Neptune 7.8 and Pluto a very faint 14 mag.

THE SOLAR SYSTEM

The Solar System is taken to include everything that is bound by gravity to the Sun, the major planets, minor planets or asteroids, comets, meteoroids and interplanetary dust. The orbits of the planets from Mercury to Neptune are shown on the inside front cover. The whole of the upper diagram lies well within the orbit of Jupiter. The main belt of asteroids lies between Mars and Jupiter but some pass inside the Earth's orbit. Some comets remain within the orbits of the major planets while others in elongated orbits pass far beyond Neptune. There are probably millions of comets and other icy bodies, of which Pluto may be one, in this remote region of the Solar System. Meteoroids are the debris of comets and asteroids when in space, some of which are seen as meteors as they burn up in our atmosphere. If they reach the ground they are called meteorites (see pp. 1 & 2 for an explanation of terms used).

The outer planets move more slowly than the Earth and have larger orbits, taking over a year to orbit the Sun. Mars takes 1.9 years, Jupiter 12, Saturn 29.5, Uranus 84, Neptune 165 and Pluto 248 years. So, for example, Saturn will take over twice as long as Jupiter to pass through a given constellation.

Although the Sun, Moon and planets all move eastwards (direct motion) against the background of the stars, the outer planets retrograde or move westwards for a few weeks at about the time of opposition, as they are overtaken by the faster-moving Earth. The inner planets Mercury and Venus can also retrograde near inferior conjunction when they are closest to the Earth. This general eastwards motion is not to be confused with the daily E–W movement of the whole sky due to the Earth's rotation.

While Mars is making one orbit of the Sun, the Earth will have made nearly two and it will be 2 years and 2 months before they line up again with the Sun and Mars comes to opposition. After opposition Mars remains in the evening sky for many months, setting about the same interval after sunset, until eventually it is overtaken by evening twilight. So we get to see Mars well every 2 years for a few weeks, and at the closest oppositions it can equal Jupiter in brightness. But due to its eccentric orbit its opposition distance from the Earth varies between 56 million and 100 million km over a 15 year cycle, so its peak brightness varies between about -2.8 and -1.0 mag.

The inner planets Mercury and Venus also move eastwards with the Sun but oscillate between E and W of the Sun as they become evening or morning objects. Mercury takes 88 days to orbit the Sun and Venus 225 days, but being in orbits smaller than the Earth's they can never appear far from the Sun (see inside cover). The greatest possible elongations are 28° for Mercury and 47° for Venus. Mercury has three morning and three evening apparitions in an average year but is usually seen easily with the naked eye from our latitude at only two of these. Venus takes 19 months from one superior conjunction to the next, and like Mercury not all apparitions are equally favourable.

It is our changing viewpoint on the Earth as it orbits the Sun that causes the Sun to appear to move eastwards day by day bringing into view different stars at different seasons. The Sun's apparent path round the sky is called the ecliptic and it is inclined at 23.4° to the celestial equator. To an observer on the Sun it would also be the Earth's apparent path. The Earth would be where the Sun was 6 months earlier. The Moon and planets are always found near the ecliptic, though each can wander a few degrees above or below it as their orbits are not quite in the same plane as the Earth's.

Further events in 2006

The Moon's orbit is inclined by 5.2° to the ecliptic so it can swing as much as 28.6° above or below the celestial equator, a complete cycle taking 18.6 years. The angular distance of a body above or below the equator is known as its declination (N or S).

The Moon reaches these extreme declinations this year. The effect will be most noticeable in two ways, the height above the horizon as the Moon crosses the meridian (is in transit and due south) and the points on the horizon (azimuths) where the Moon rises and sets. Both will depend on your latitude. The farther north you are the less the altitude at transit but the farther round towards the N the Moon will rise and set when at high N declination.

The Moon often reaches extreme N and S declinations in the same month, but this year the most southerly is on 22 March and the most northerly on 15 September. Each month in 2006 the Moon will reach N and S extremes almost as great as in these peak months. On 22 March the Moon will reach -28.6 ° S declination (measured from Earth's centre). In London the Moon will be above the horizon for only 6.5h. It will be at last quarter and due S at 05h 37m, just 10° above the S horizon.

On 15 September the Moon will reach +28.6° above the equator, reach a peak altitude of 67° and be above the horizon for 17h 20m in London. It will be at last quarter and due S at 06h 34m.

Perhaps these extremes are most noticeable around full Moon. The Moon will be near full when near greatest N declination on 12 January and 6 December, and near greatest S declination on 13 June, 10 July and 7 August.

From 9th to 27th of August Mercury, Venus and Saturn will be close in the NE dawn sky from about 04h 30m (see August notes). They may be observable in the bright sky. The crescent Moon will join them over 21st–22nd.

In early December Mercury, Mars and Jupiter will be close together in the SE sky about dawn. Only a few degrees up at 07h they should still be visible to the naked eye. On the 7th Mercury will be above Mars, with Mars above Jupiter. By the 10th all three will be within a circle about twice the Moon's diameter or 1°across. By the 13th Jupiter will be above Mars, with Mars above Mercury. It is very unusual to have three planets so close together. They should all be easily seen in binoculars even in a bright sky. Jupiter is much the brightest at -1.7 mag, then Mercury at -0.6 mag and the faintest Mars at +1.5 mag.